AF265766

4th Grade Math
Volume 5

Table of Contents

Angles

Key Vocabulary

acute angle

obtuse angle

right angle

Describe a right angle.

A right angle is an angle that has a measure of 90°

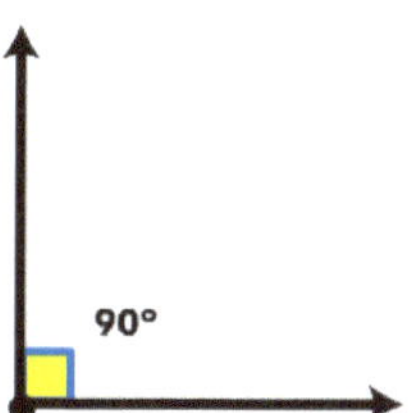

List 5 things around you that have right angles.

1.____

2.____

3____

4.____

5.____

Here are some other types of angles.

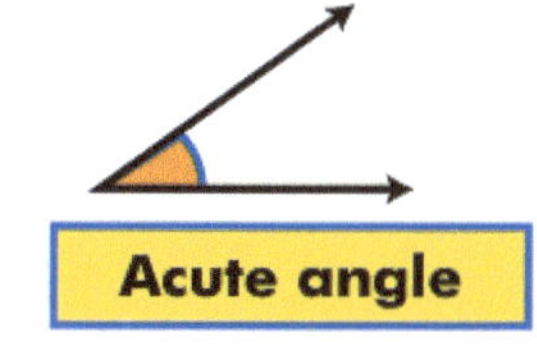

An angle with a measure of less than 90°.

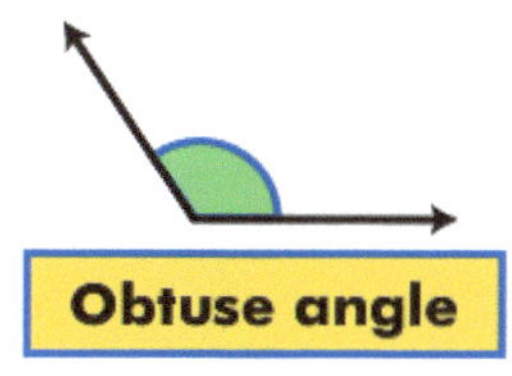

An angle with a measure of greater than 90°, but less than 180°.

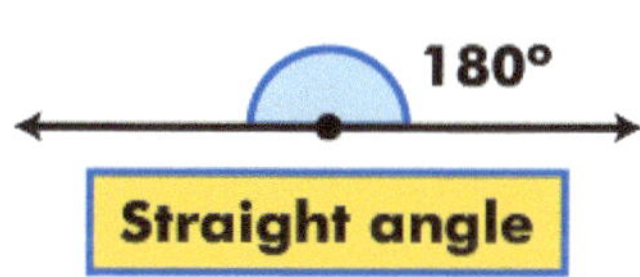

An angle with a measure of 180°.

Label these angles.

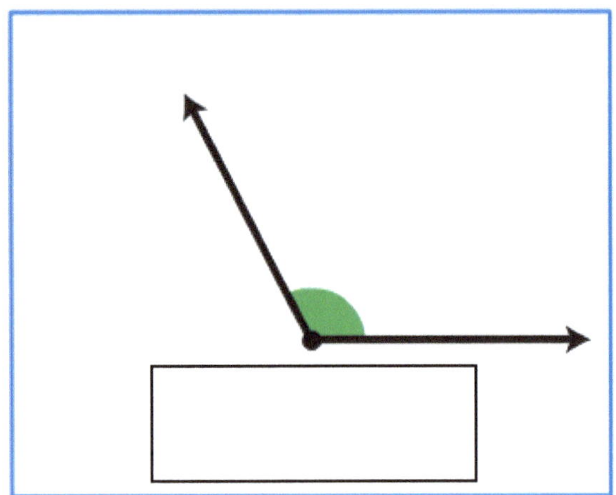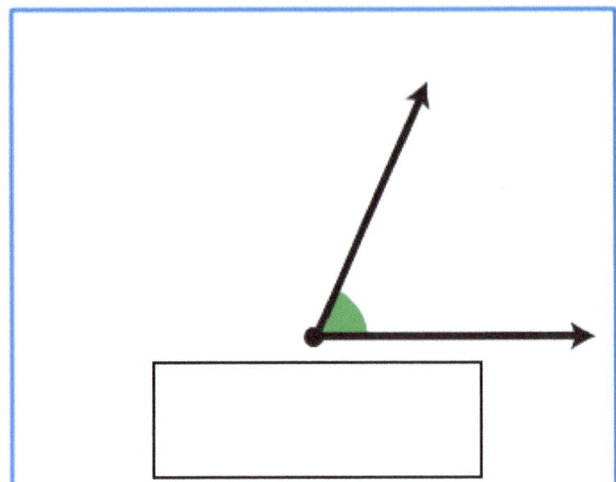

Label these angles.
Use the letters in the circles for your labels.

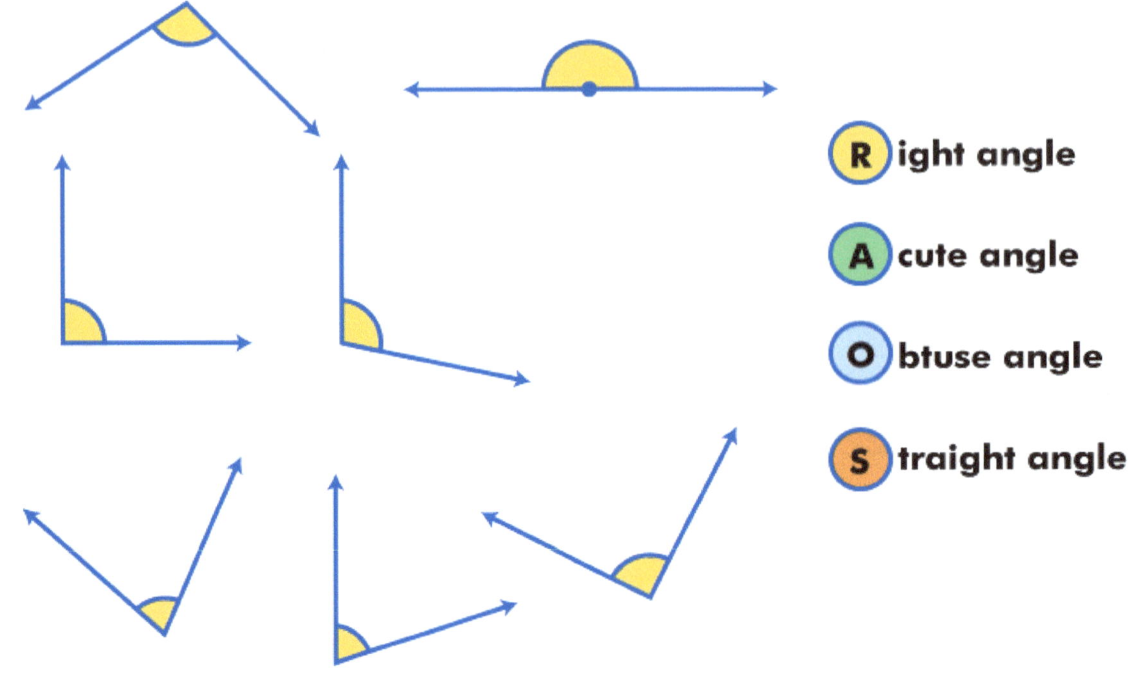

Estimate the measure of these angles.

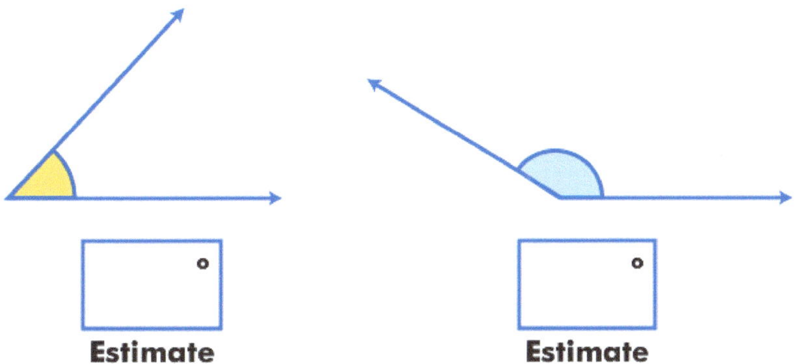

Estimate   Estimate

A right angle gives you the information that you need to estimate the measurement.

USING A PROTRACTOR

Make sure that you position the protractor correctly (Figure 1).

Always read up from zero using either the outside scale (Figure 2) or the inside scale (Figure 3).

Read the protractor to determine the measure of the angle (Figure 4).

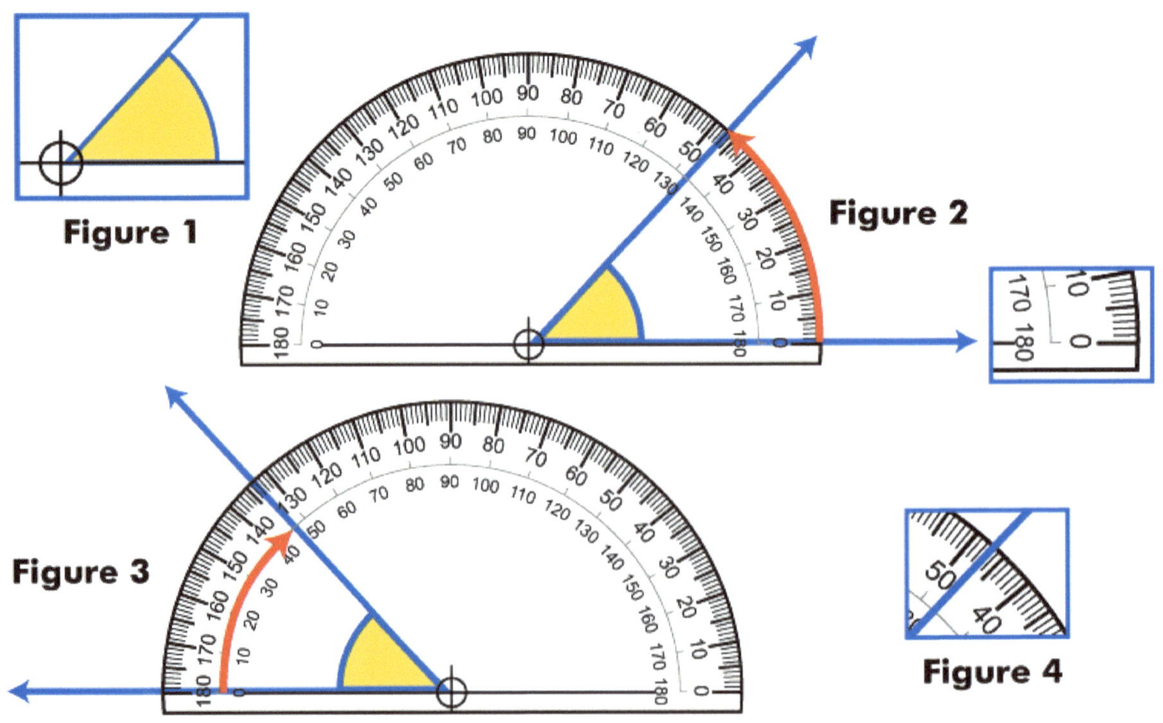

Figure 1

Figure 2

Figure 3

Figure 4

Name______________________________

Angles Quiz

Fill in or circle the correct answer.

1   **True or false? An obtuse angle is greater than a right angle.**

2   **What is the measure of a straight angle?**

- **A**   **90°**
- **B**   **45°**
- **C**   **100°**
- **D**   **180°**

3   **What is the measure of the obtuse angle (in degrees)?**

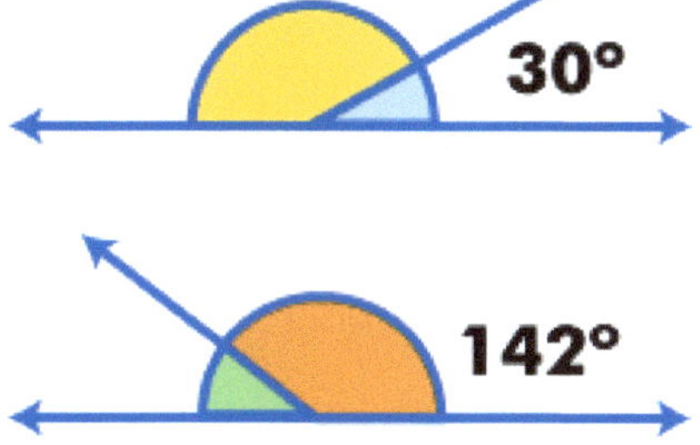

4   **What is the measure of the acute angle (in degrees)?**

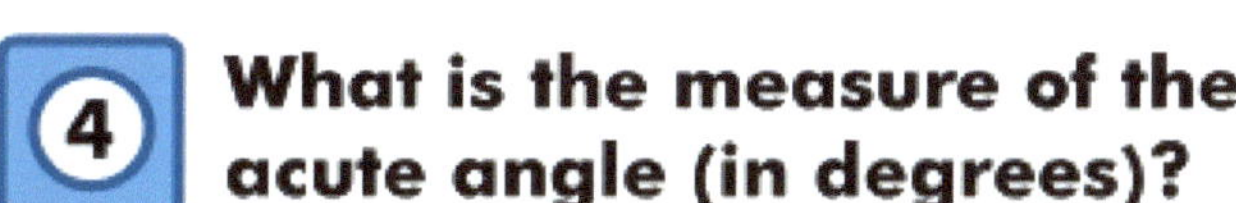

Transformations

Key Vocabulary

rotate

reflect

translate

congruent

mirror line

transformation

Are these shapes congruent?

Congruent: figures that are the same shape and size.

How can you be sure that these shapes are congruent or not? ________________

____

____

____

Describe the movement to superimpose one figure over another.

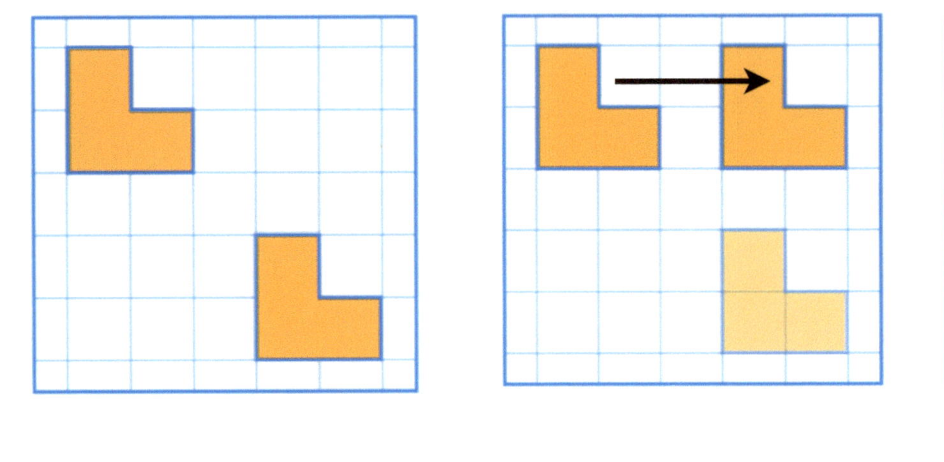

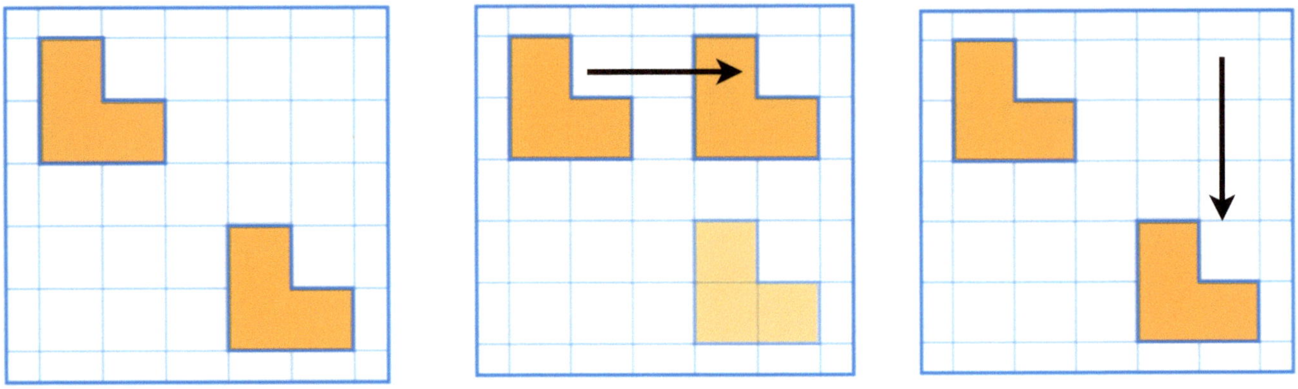

Step 1 Move _______ spaces to the _____________.

Step 2 Move ____.

We call this *a translation or a slide.*

> ## We call this *a reflection* or *a flip*.

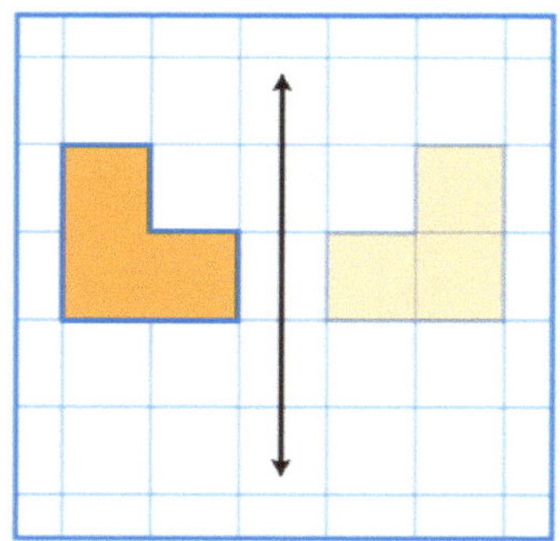 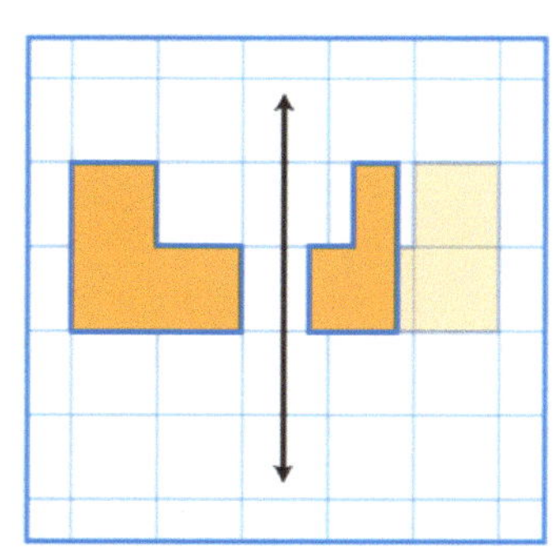 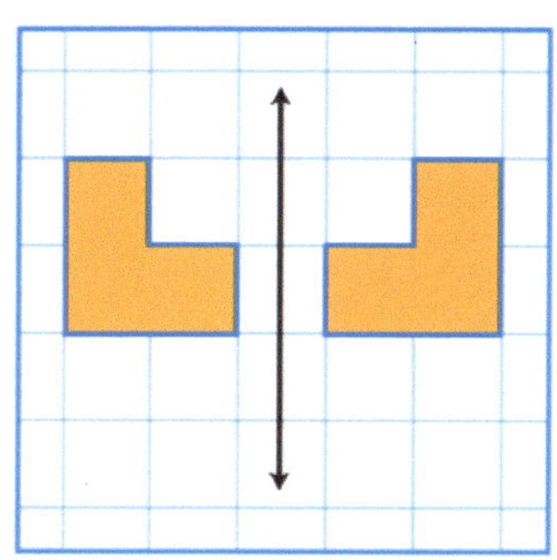

Think about turning a page in a book. Is this a reflection or a flip? _________________

———————————

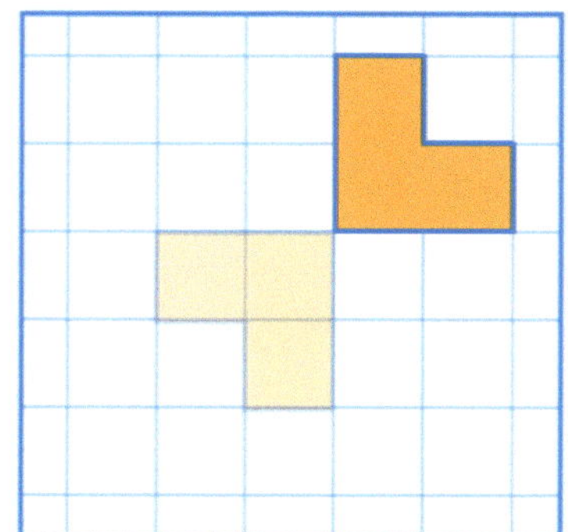 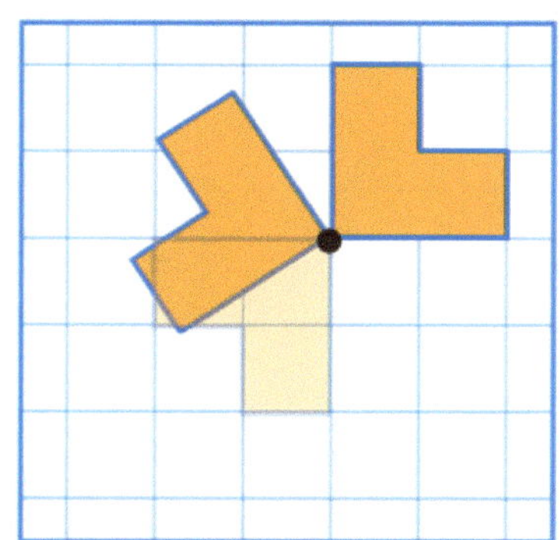 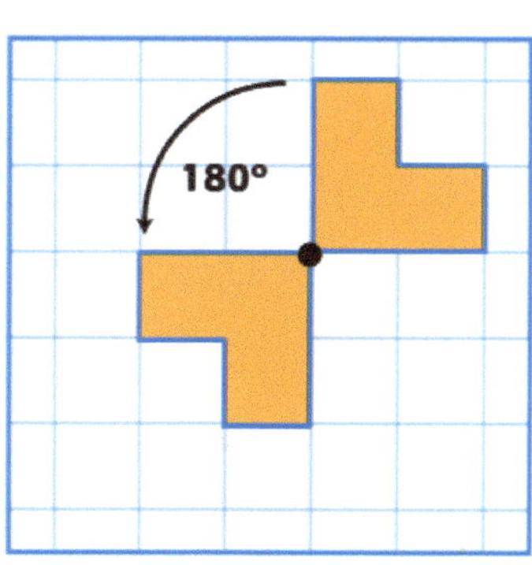

We call this *a rotation* or *a turn*. Drag
the point to the center of the rotation.

Show and describe a translation.

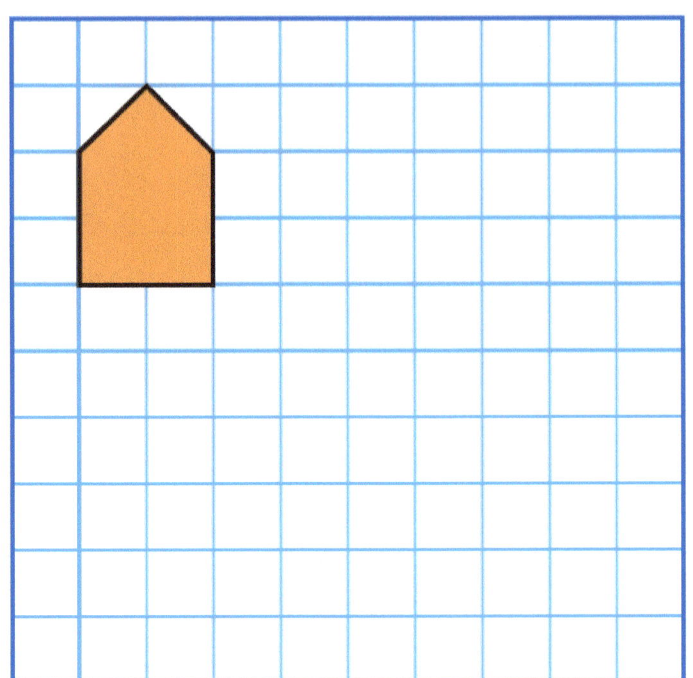

____

____

____

____

____

____

Draw a reflection.

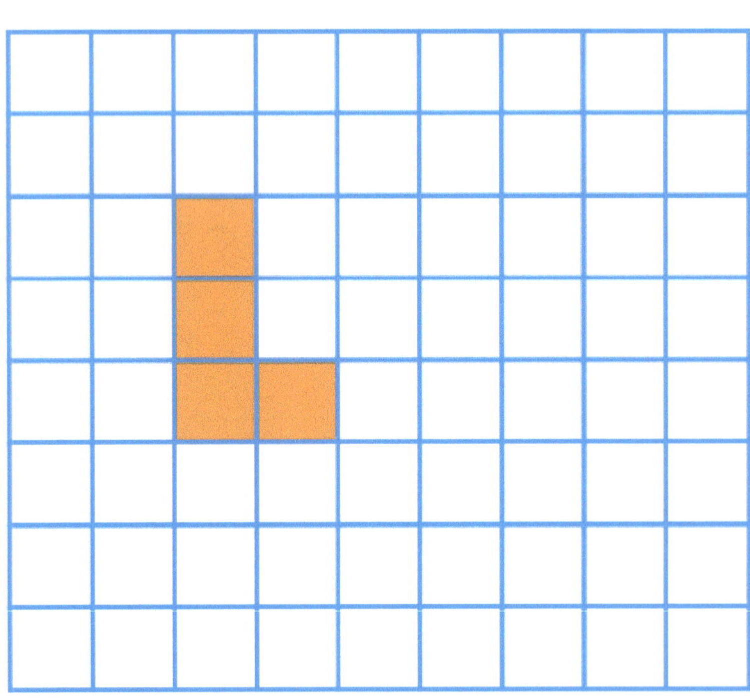

Can you show a rotation using this graph. Use the point as the center for the rotation.

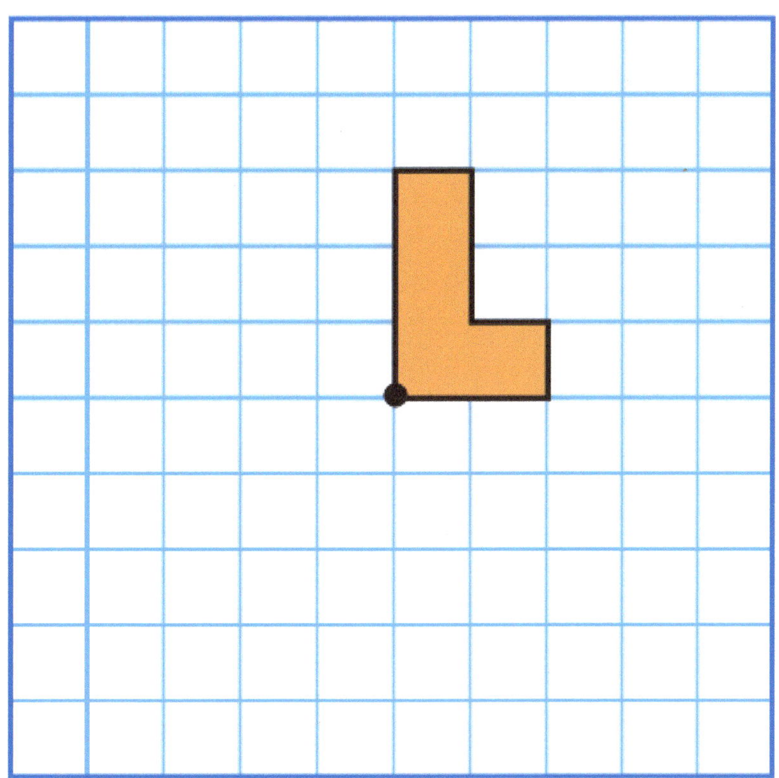

Label the transformations.

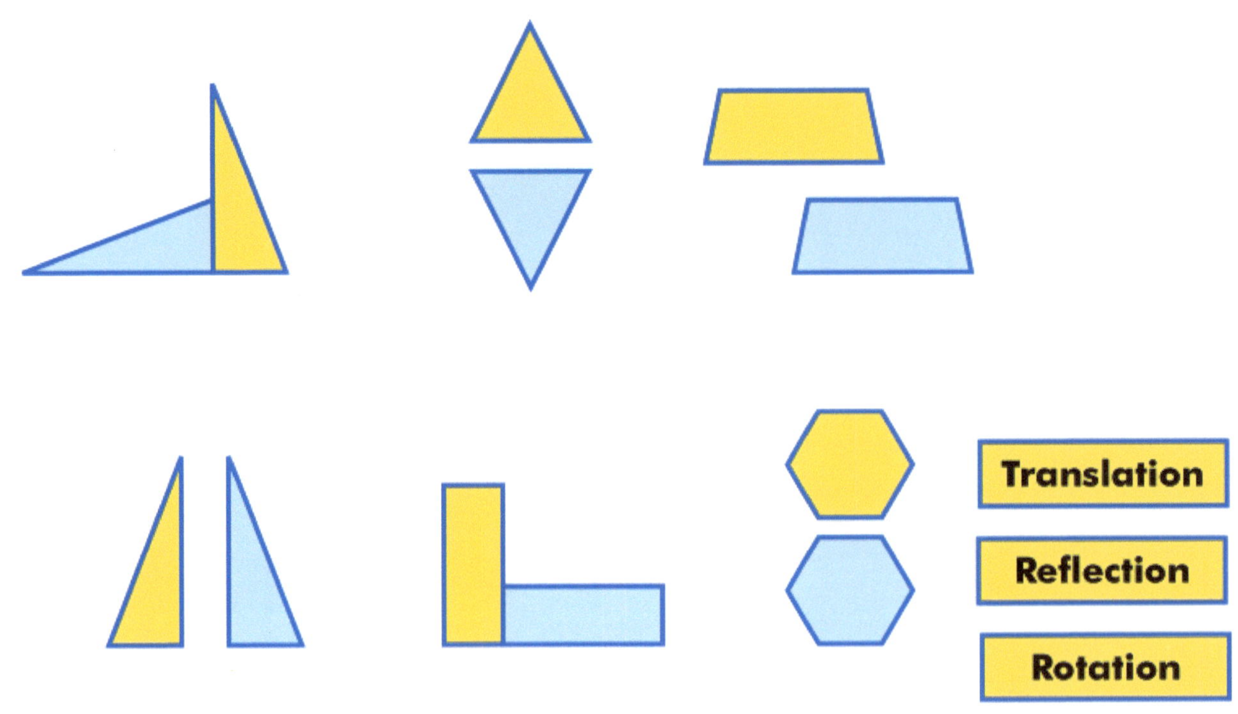

How many degrees.

How many degrees did each figure rotate when moving clockwise and counter-clockwise. The yellow figure is the starting point.

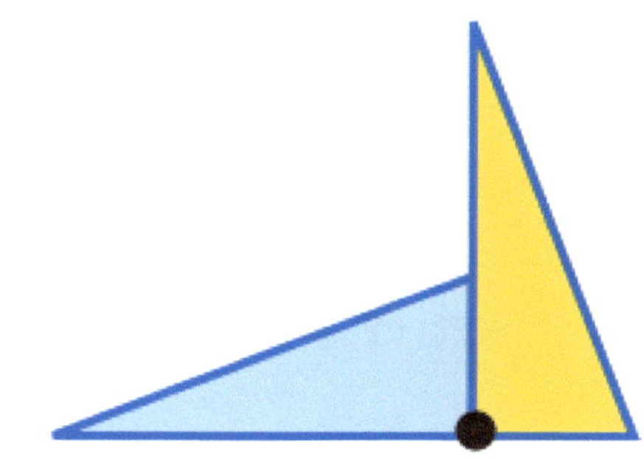

Clockwise __________

Counter Clockwise __________

Clockwise __________

Counter Clockwise __________

Name_________________________________

Transformation Quiz

Answer yes or no in the gold box for each figure and label.

Decide if each transformation has been correctly described.

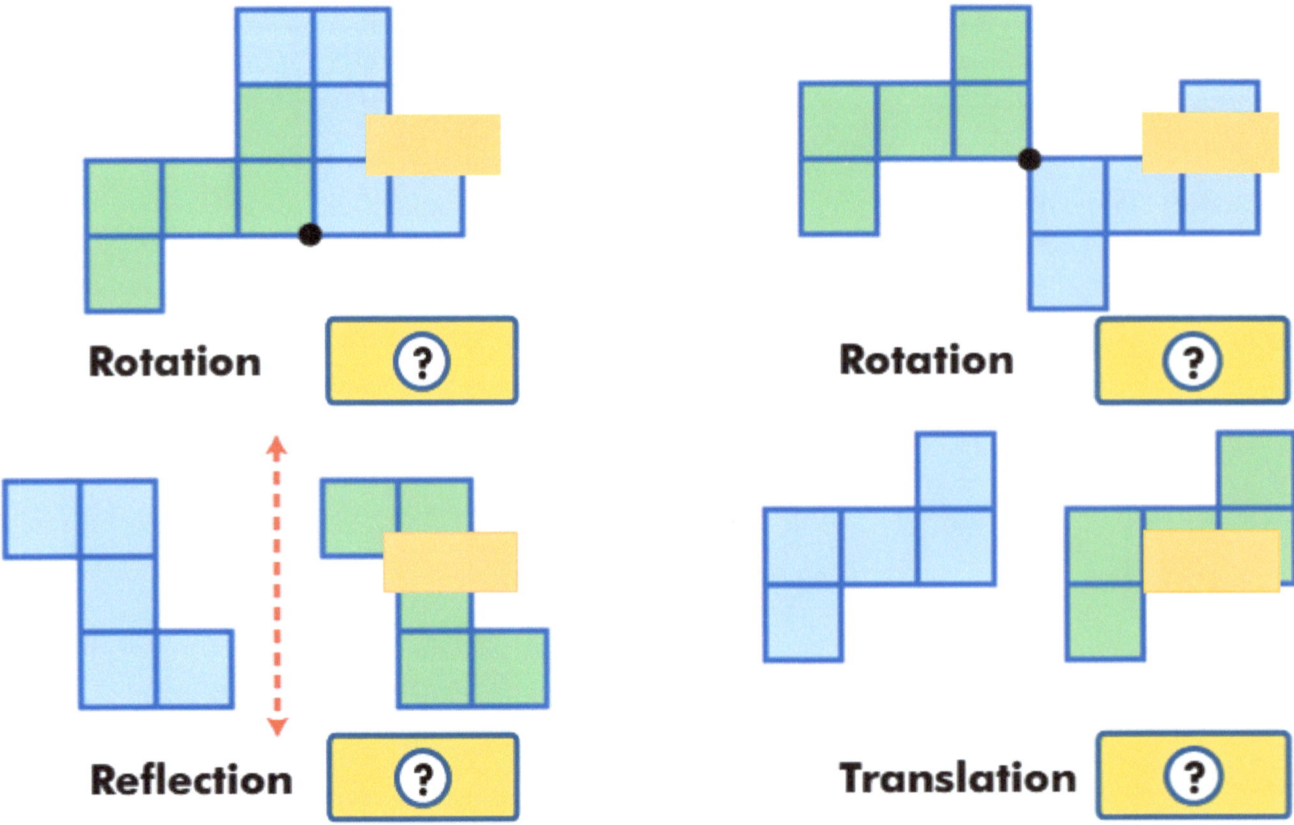

Units of Measure

Key Vocabulary

length

inch

foot

yard

mile

Order these customary units of measure.

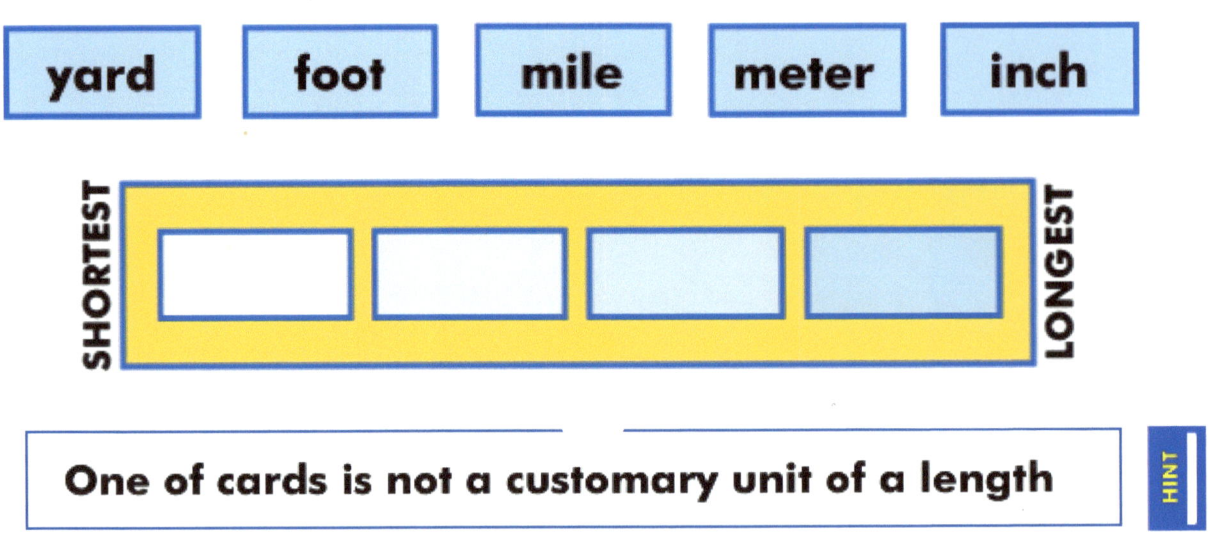

One of cards is not a customary unit of a length

HINT

____

Find the lengths.

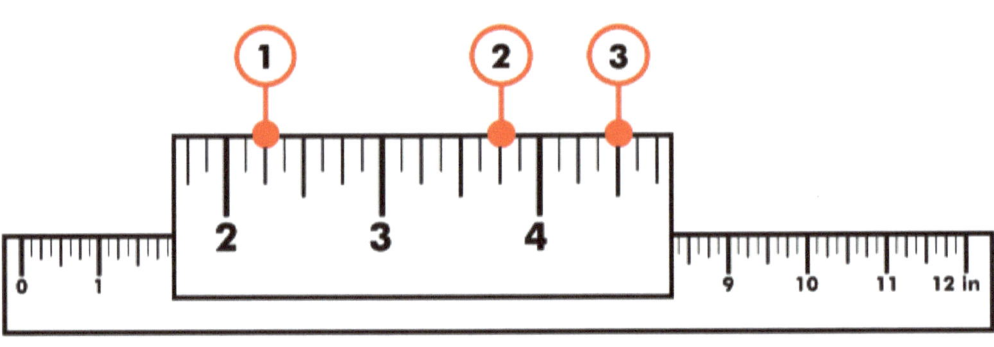

(1) _____________ inches

(2) _____________ inches

(3) _____________ inches

More Practice Finding Lengths
Write the length above the ruler.

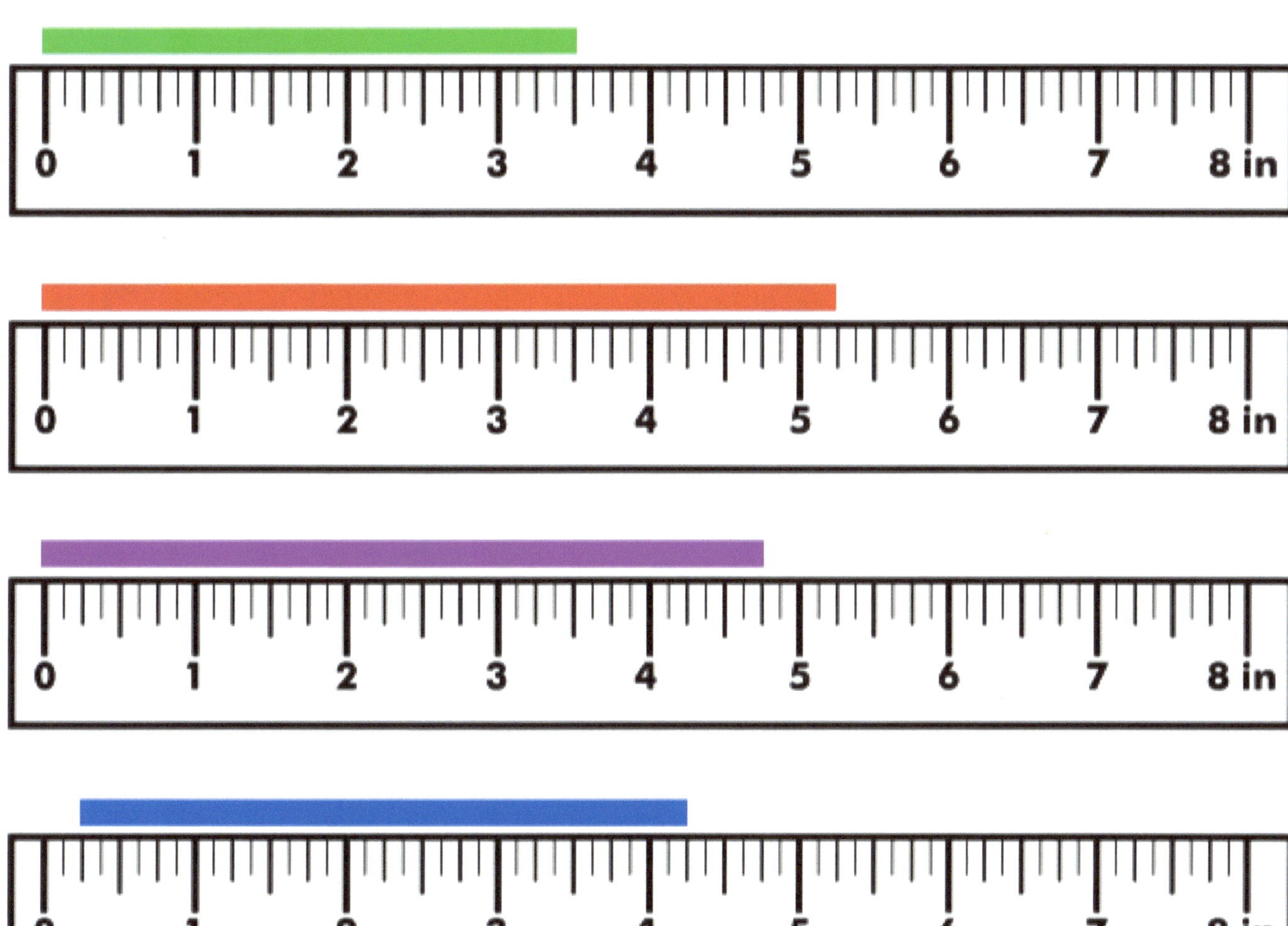

Connect the equivalent lengths with a line.

1 ft	36 in
1 yd	5,280 ft
1 yd	1,760 yd
1 mi	12 in
1 mi	3 ft

Changing Units of Length.
Use the key to help convert the units of length from one unit to another.

4 yd = [   ] ft

10 mi = [   ] yd

15 ft = [   ] yd

72 in = [   ] yd

Key

1 ft	=	12 in
1 yd	=	36 in
1 yd	=	3 ft
1 mi	=	5,280 ft
1 mi	=	1,760 yd

Name________________________________

Units of Measure Quiz
Circle or fill in the correct answer.

1 **True or false? 5 ft = 62 in**

2 **The Eiffel Tower is 1,986 ft tall. How many yards is this?**

- **A** 5,958
- **B** 662
- **C** 1,665
- **D** 23,832

3 **How many yards are there in 1 mile?**

4 **How many feet are there in 1 mile?**

Perimeter

Key Vocabulary

perimeter

area

square units

irregular figure

Perimeter of a Rectangle

Start at the point and draw a line along the black line that outlines the rectangle. How many units (boxes) do you cross? That is the perimeter of this rectangle.

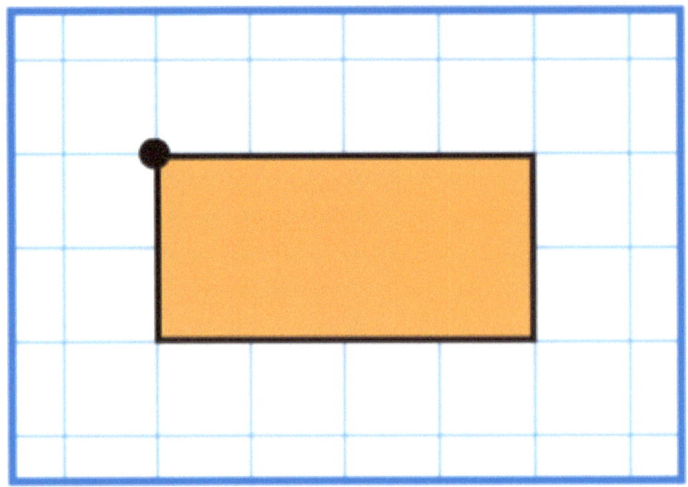

What is the perimeter of this rectangle? ____________ units.

The area of the rectangle includes the entire rectangle not just the perimeter.

What is the area of this rectangle?

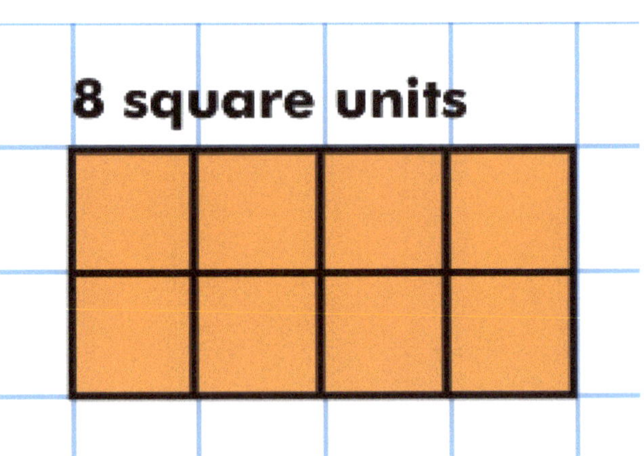

Find the area and perimeter for each figure.
Connect the answer with the correct figure.

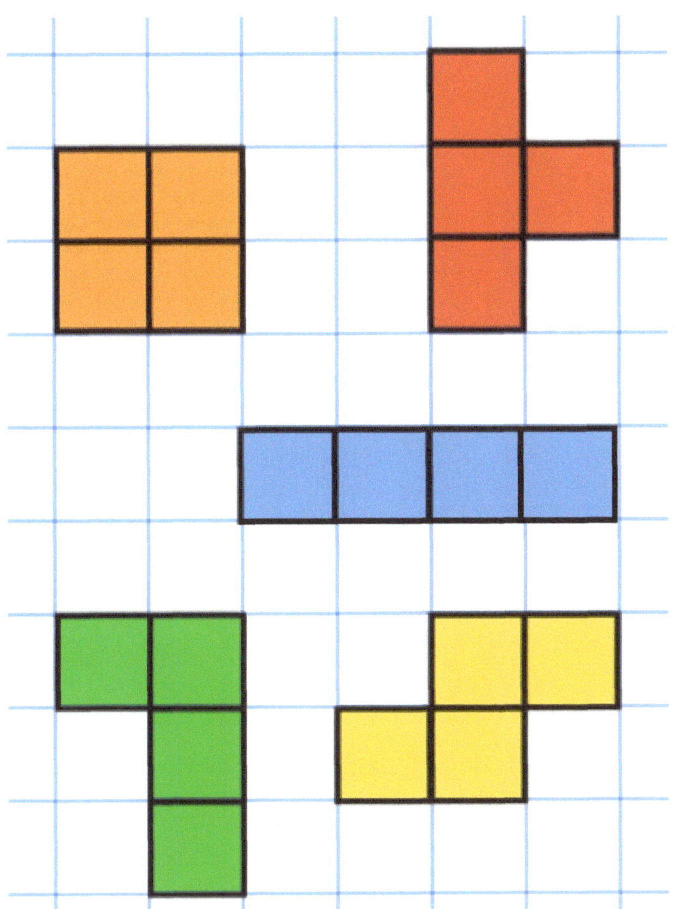

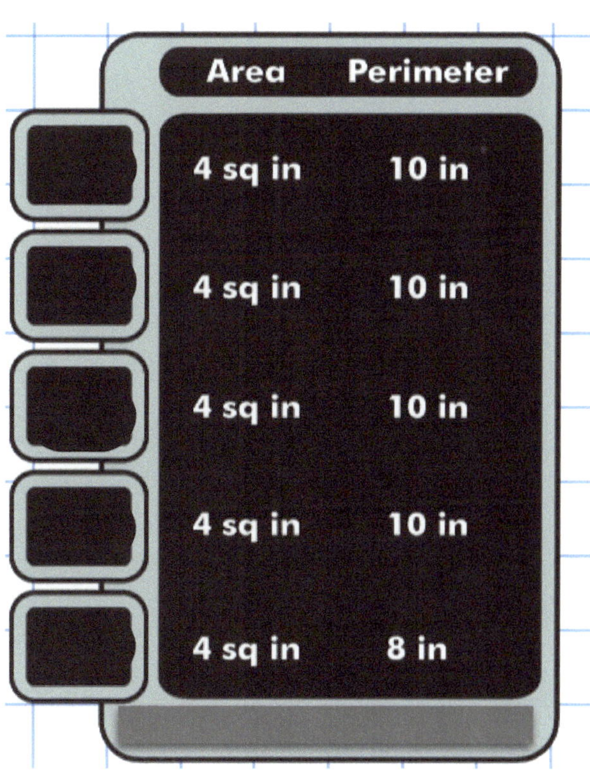

Practice Square Perimeters

Fill in the blanks

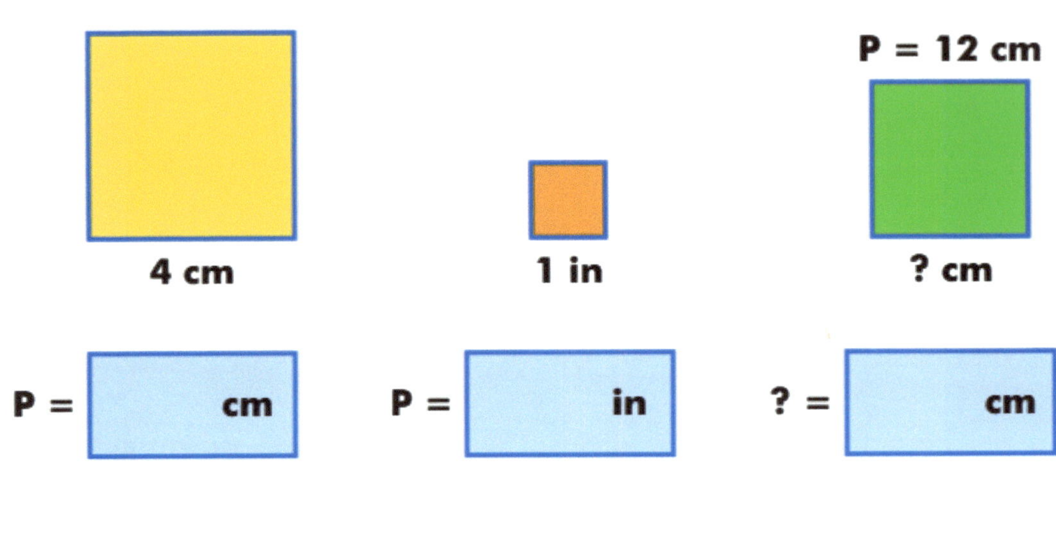

Find the perimeters to these regular polygons and label it with one of the suggestions below.

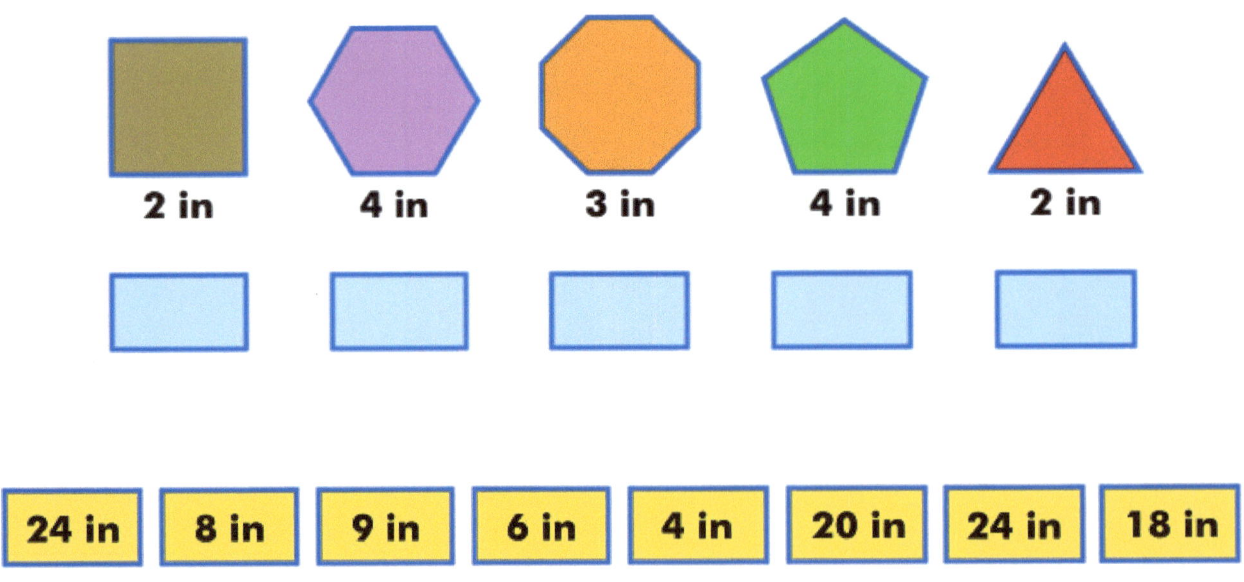

Find the perimeter of this irregular figure.

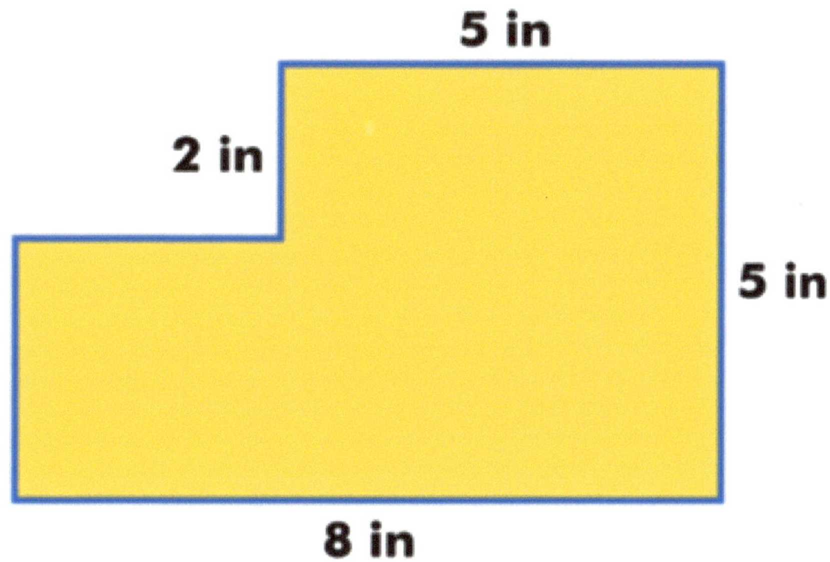

Name_________________________________

Perimeter Quiz
Circle or fill in the correct answer.

1 **True or false? Two rectangles with the same area always have the same perimeter.**

2 **Which figure has the smallest perimeter?**

A

B

C

D

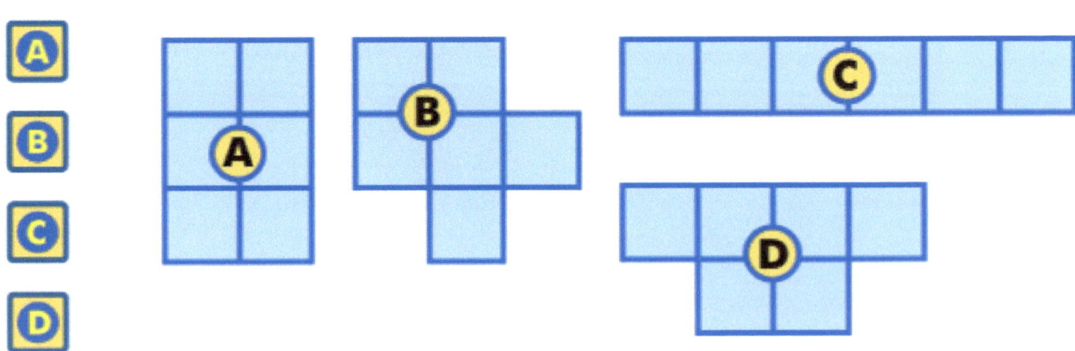

3 **What is the perimeter of the figure with the largest perimeter (in units)?**

4 **What is the length in cm of one of the sides of a regular pentagon with a perimeter of 17.5 cm?**

Area

Key Vocabulary

area

square units

Find the Area

How can you find the area of this rug?

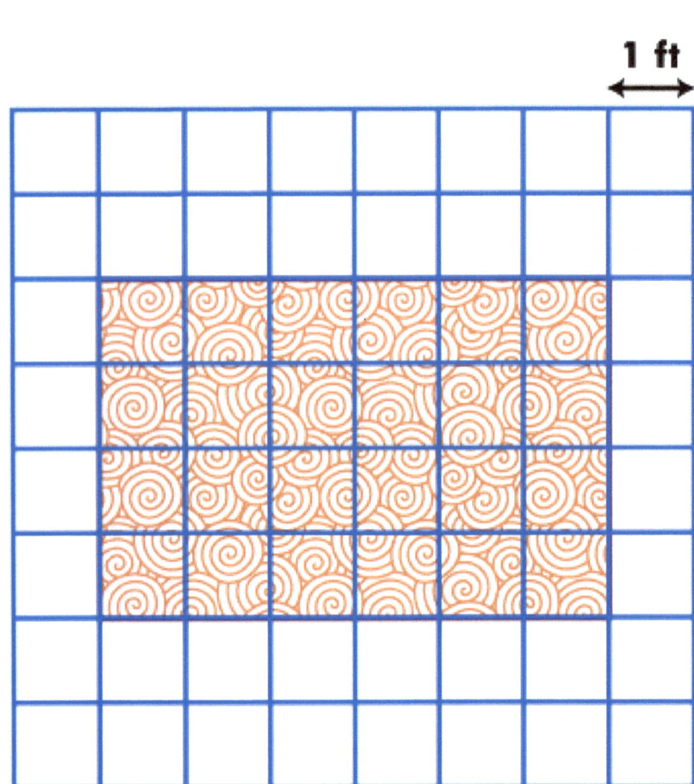

We could use a grid and count the number of squares (24). The area of the rug is 24 sq ft.

A Different Method to find the Area

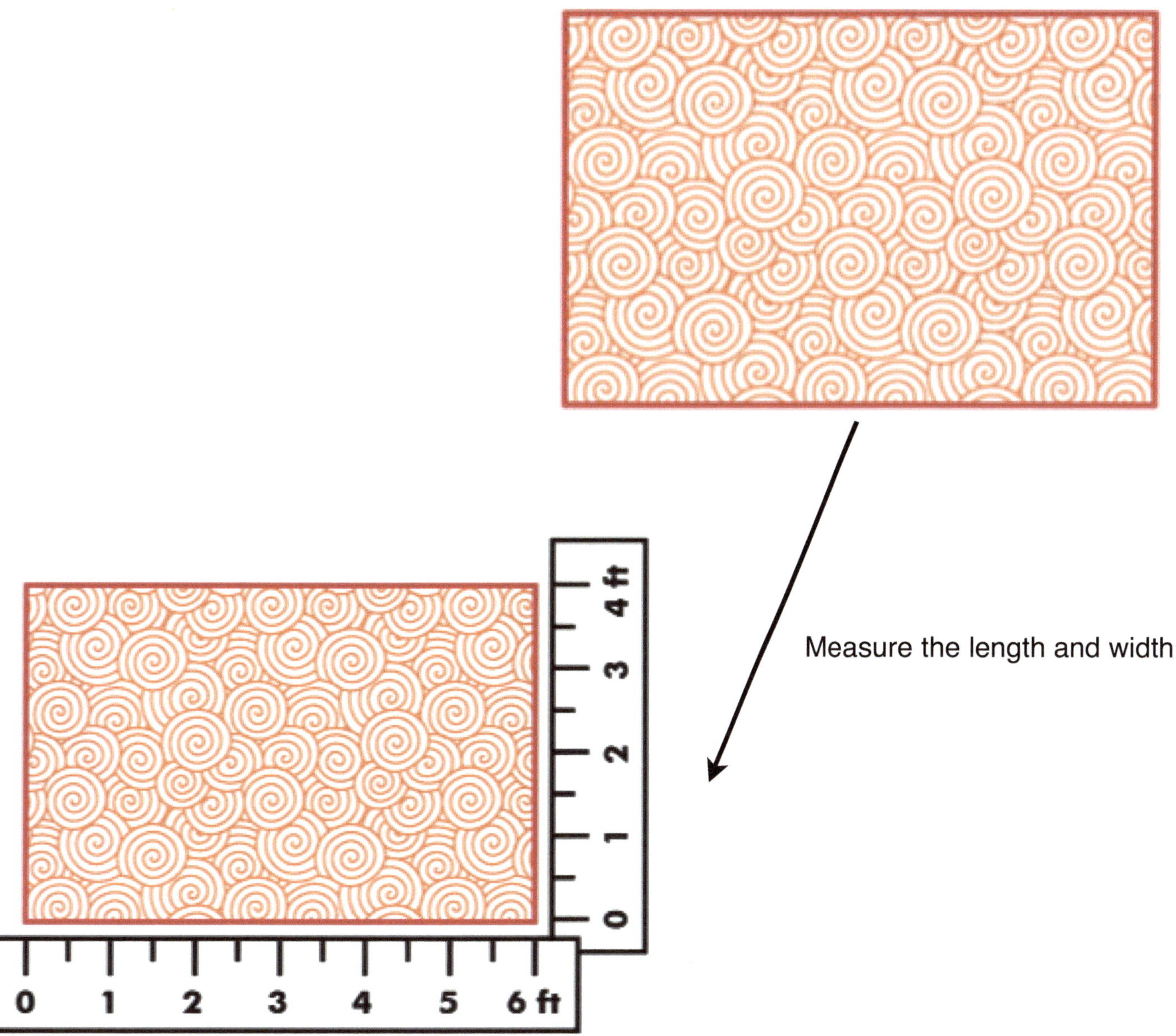

Measure the length and width

Area = l x w

Area = 6 ft x 4 ft

Area = 24 sq ft

Area of a rectangle = length x width

Match the rectangles that have the same area.
Place the matching rectangles letters side by side on the table and then add the ft².

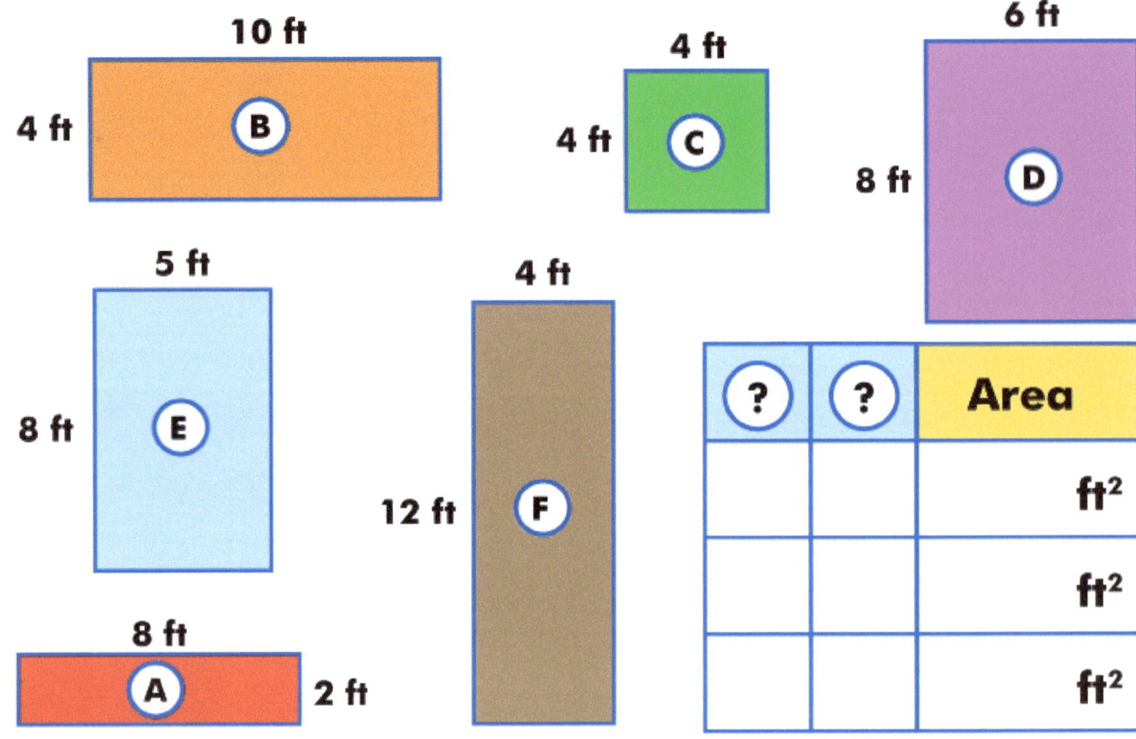

Area of a Right Triangle
Can you solve for the area of the blue triangle?

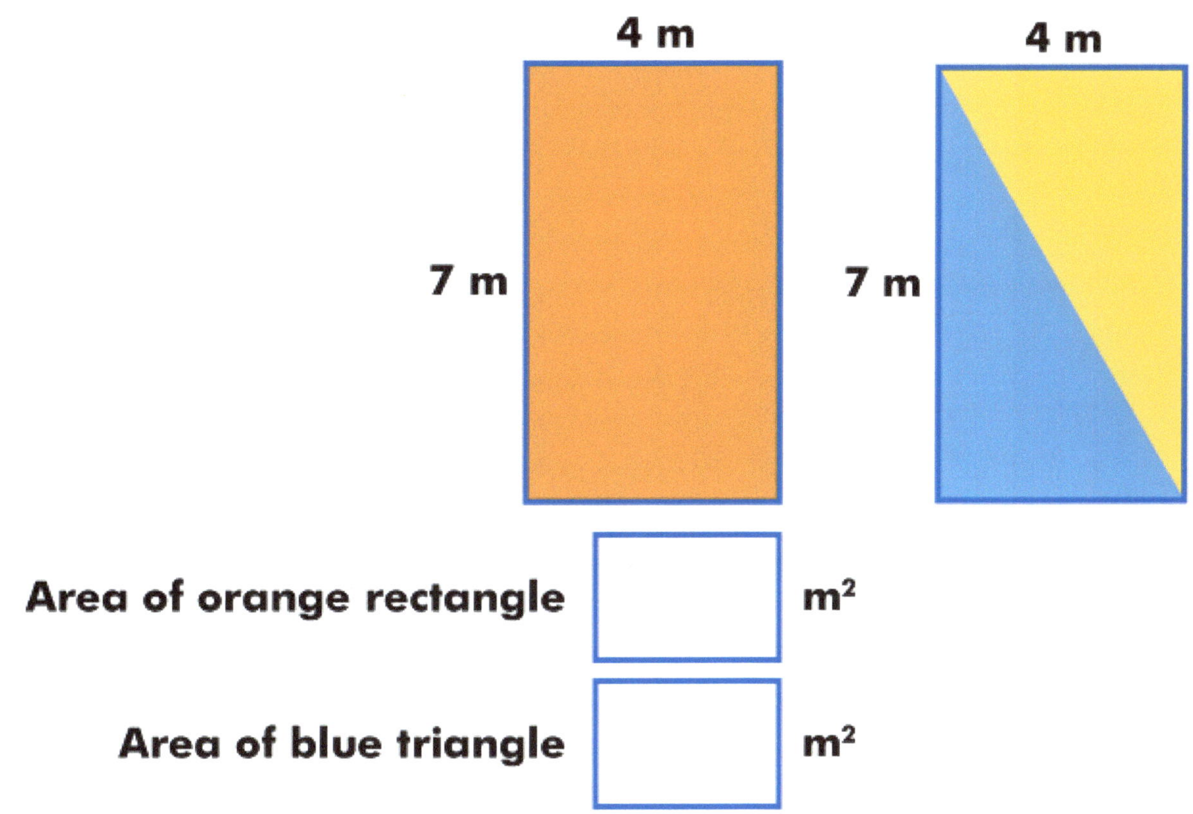

Area of orange rectangle [ ] m²

Area of blue triangle [ ] m²

Find the area of the blue triangles.

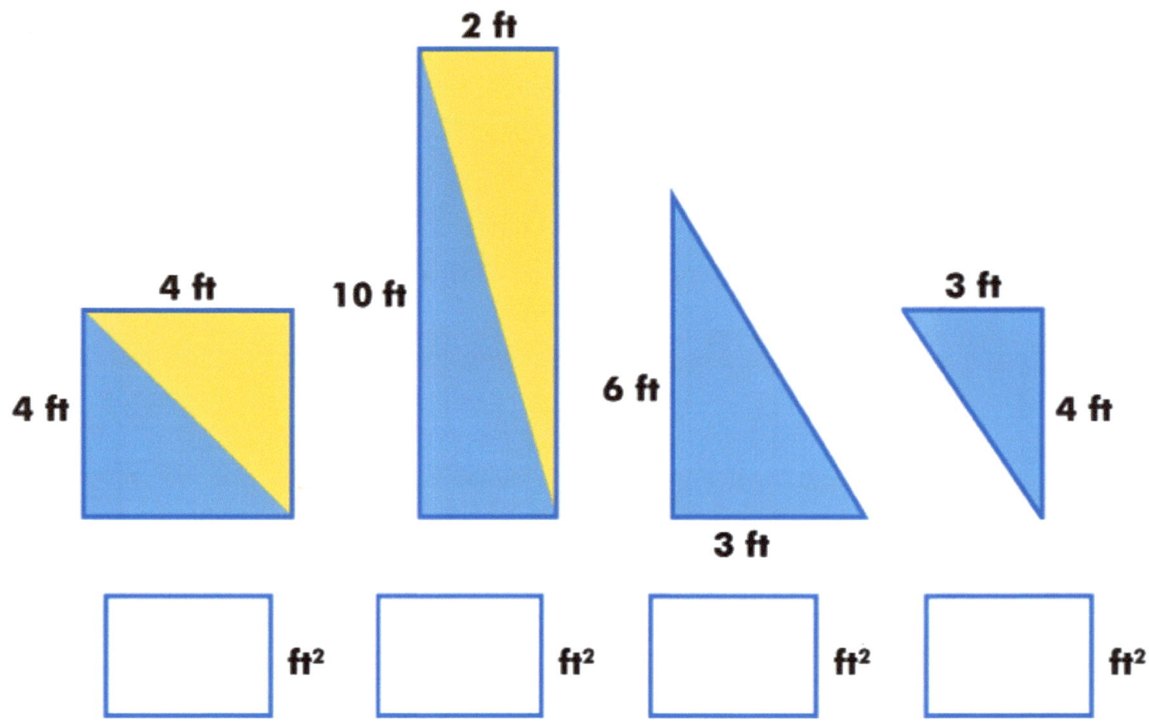

Area of irregular figures.

Find the area of this irregular figure. There is a color coded hint below.

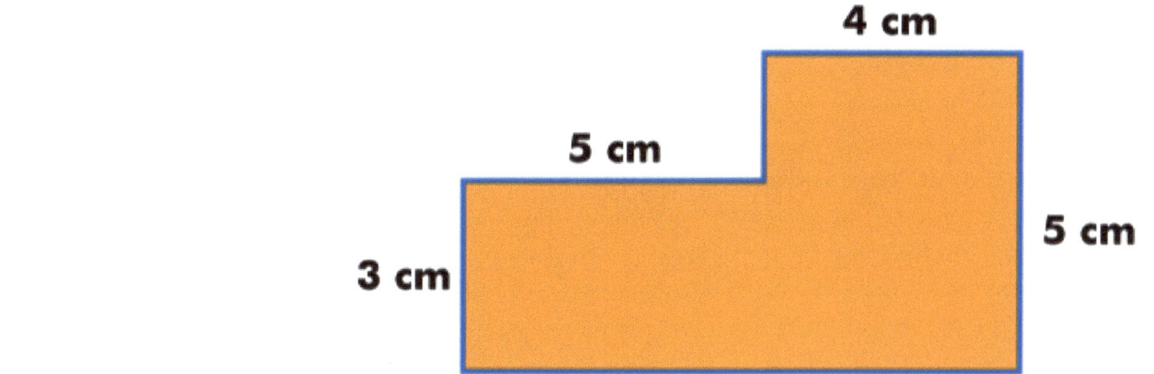

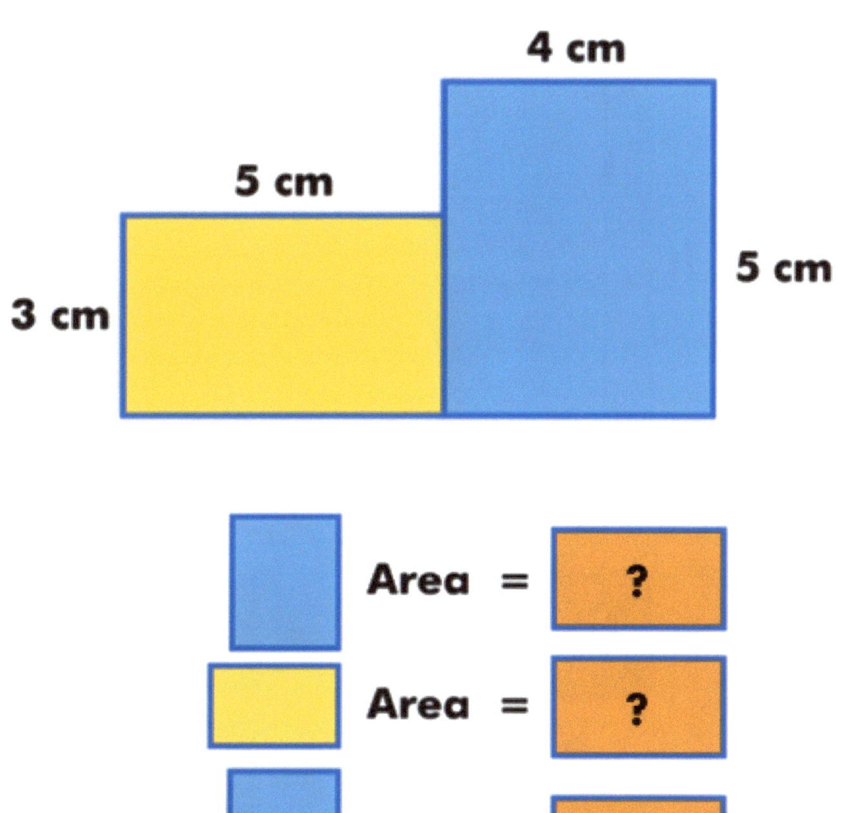

Another Irregular Figure

Find the area of this irregular figure. The same color coded hint is provided to help you solve.

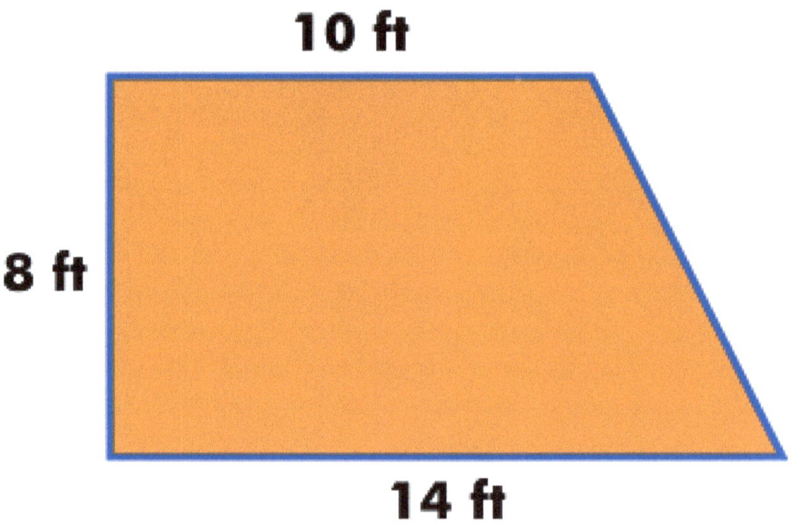

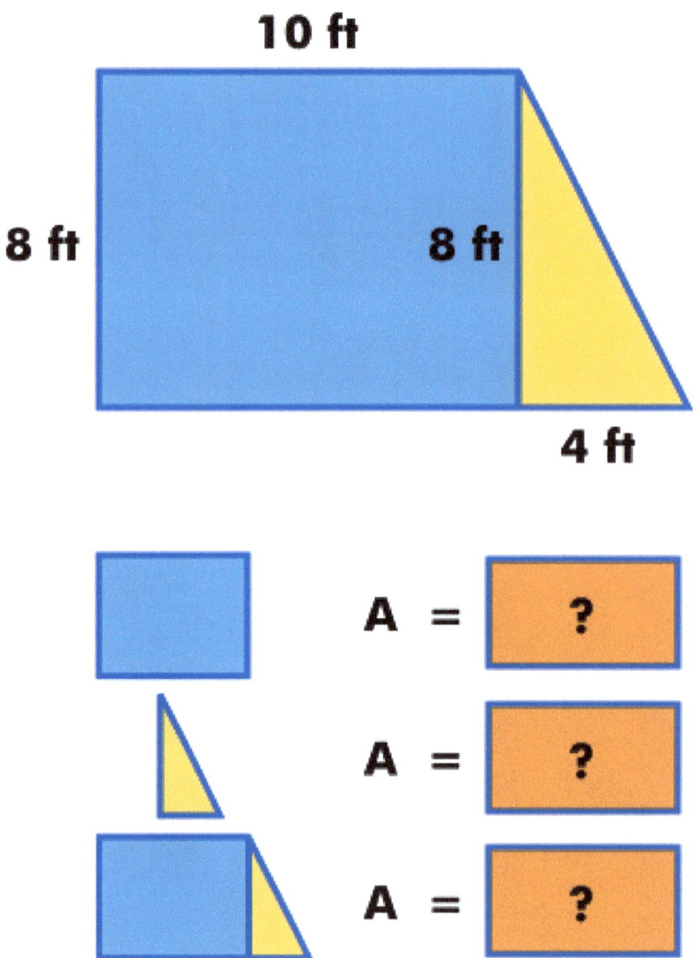

Real Life Example

How much carpet is needed for this room?

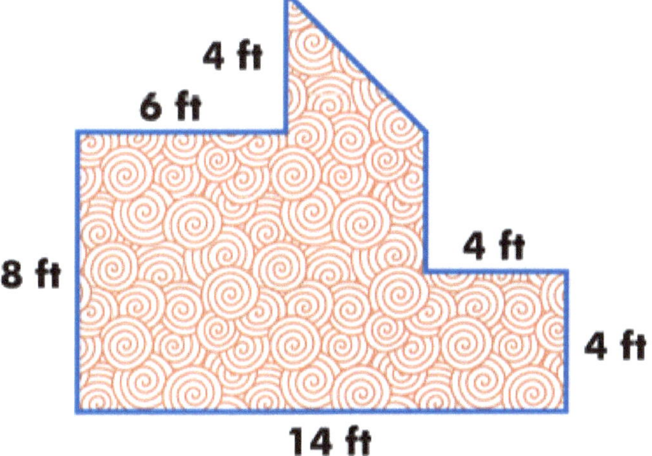

Hint

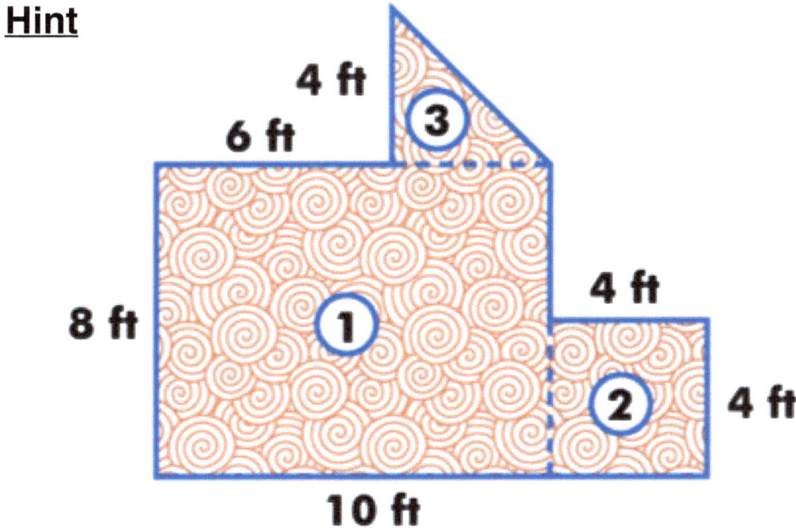

① **A =** ?

② **A =** ?

③ **A =** ?

A =

Name_________________________________

Area Quiz

1 True or false? The area of a triangle is half the area of a rectangle with the same base and height.

2 What is the area of the blue triangle in Figure 1?

A. $15\ ft^2$

B. $7\frac{1}{2}\ ft^2$

C. $53\ ft^2$

D. $35\ ft^2$

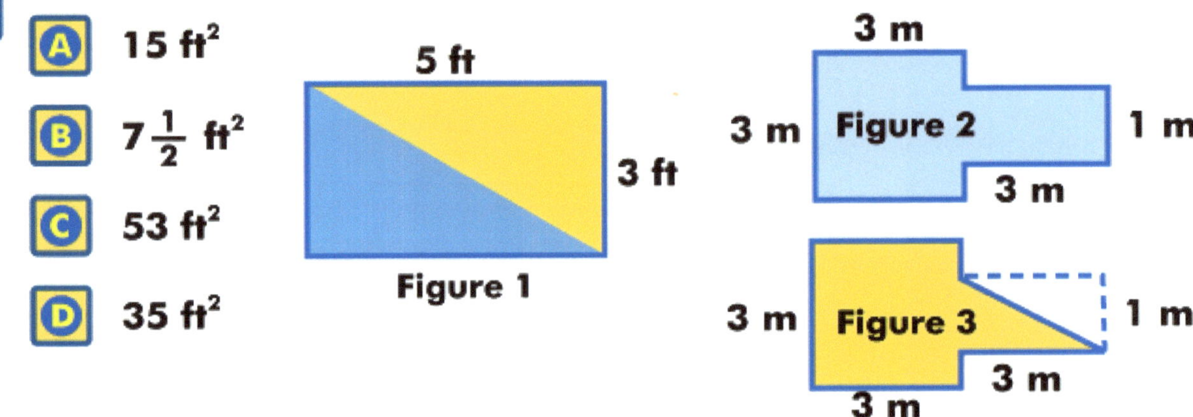

3 Find the area of Figure 2 in square meters.

4 Find the area of Figure 3 in square meters. Give your answer as a decimal.

Newburyport, MA 01950

1-800-596-3175

OnBoard Academics employs teachers to make lessons for teachers! We create and publish a wide range of aligned lessons in math, science and ELA for use on most EdTech devices including whiteboard, tablets, computers and pdfs for printing.

All of our lessons are aligned to the common core, the Next Generation Science Standards and all state standards.

If you like our products please visit our website for information on individual lessons, teachers licenses, building licenses, district licenses and subscriptions.

Thank you for using OnBoard Academic products.